VOLCANO!!

Text by
Scott C.S. Stone

AN ISLAND HERITAGE BOOK

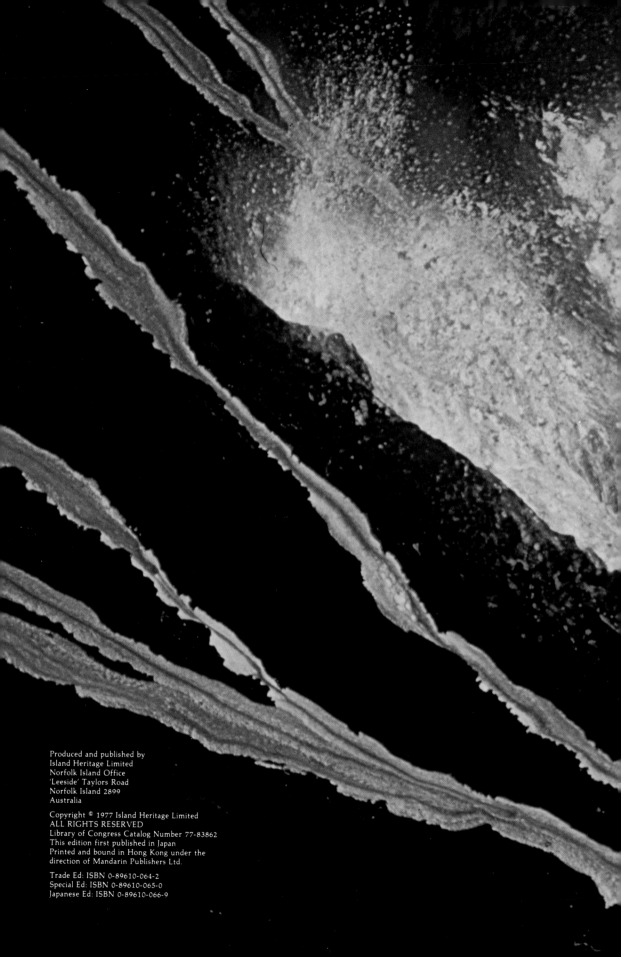

Produced and published by
Island Heritage Limited
Norfolk Island Office
'Leeside' Taylors Road
Norfolk Island 2899
Australia

Copyright © 1977 Island Heritage Limited
ALL RIGHTS RESERVED
Library of Congress Catalog Number 77-83862
This edition first published in Japan
Printed and bound in Hong Kong under the
direction of Mandarin Publishers Ltd.

Trade Ed: ISBN 0-89610-064-2
Special Ed: ISBN 0-89610-065-0
Japanese Ed: ISBN 0-89610-066-9

CONTENTS

LAVA POURS from vents on the rim and sides of Pauahi Crater.

ILLUMINATED BY HER FIRES, father and child watch Kilauea Iki in eruption. At left, Puu Kia'i's spatter cone sputters and rumbles between eruptions.

E PELE E

"O Goddess of the burning stones.
Here is my gift.
Life for you. Life for me.
The flowers of fire wave gently."

She has the power to attract and hold, a goddess of unearthly beauty and earthy passions. Pele rises above the flawed and contradictory legends, above fact and folklore. In the misty world of the high volcanoes the belief in Pele comes easily and strikes some primordial molecule, and the echoes ring clear and silvery in the fern and ohia forests.

The notion of a fire goddess who stamps her foot and causes earthquakes, who strikes the ground with her magic spade, Pa'oa, and causes eruptions, is not an ominous thought in that cool, clean uplands area. Older generations look for her in the desolate places of scree and pumice. Tales abound; her face gleams in the light of the eruptions, her hair floating upward with the fiery leap of the lava fountains. In the thunder of her surging flows there is the cry of pleasure.

She is seen in the firelight of the perpetual flames of the Volcano House fireplace. Static electricity forms an image in the sky, and men look and see her profile there for an instant, only an instant, but forever remembered. The old crone walking the back roads of Puna is regarded with suspicion—it is a disguise Pele sometimes adopts to mingle with men.

"PELE'S HAIR" is made when volcanic gas blows through highly fluid lava. The result is fine filaments of glass which cover adjacent lava flows. Generally honey brown in color, the gossamer glass strands break down quickly in sun and rain.

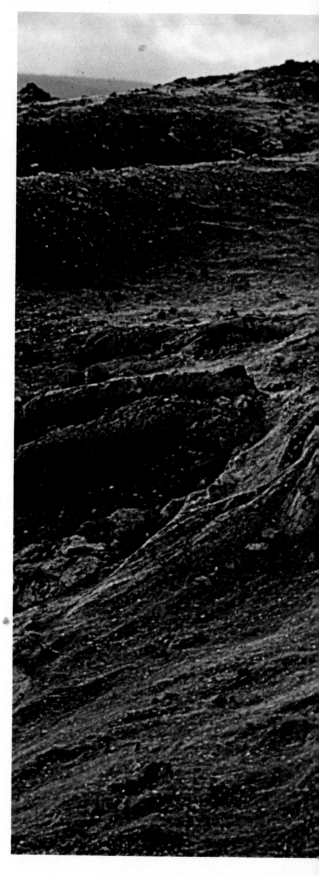

Her origins are unclear. She came from across the sea and was a wanderer. She lived on many Hawaiian islands before making her home on the largest in the archipelago, the Big Island of Hawaii. She displaced a near-forgotten fire god, the unfortunate 'Ai-la'au, who ran from Pele and so disappeared from the legends as well as from the Islands.

She had tempestuous love affairs and a love-hate relationship with Kamapua'a; she battled a snow goddess, Poli-'ahu, with whom she still shares the loftier volcanoes. She quarreled with her sisters, especially the long-suffering Hi'iaka, and sometimes left forests burnt and smoking in the wake of her anger and jealousy.

Today the older generations still offer the traditional 'ohelo berries to Pele in supplication and homage. Bemused scientists weave between a matrix of fact and legend, watching in awe when they coincide. Reporters hasten to every small eruption. Guests at the rustic old Volcano House hotel on the rim of Kilauea Caldera sip coffee and stare curiously across the caldera at the waving sulphur fumes, at the smoke rising from the clash of ground water and underground, lava-heated rocks.

In her home on Kilauea or in the great volcanic mountains of Mauna Loa and Mauna Kea, Pele is highly visible. Evidence of her wrath is everywhere in the lava flows and the striated hillsides and old

cinder cones. The rough, glittering *aa* lava reflects the swiftness of her temper, but the spectacular evening skies and the freshness of Kilauea mornings show Pele in a softer mood.

When she takes human form she often becomes a wrinkled old hag, hobbling the byways or sitting under a kamani tree, a strange old woman with blazing eyes. Motorists, especially, must be wary. The old woman hitchhiking from Opihikao to Pahoa may be just an old woman, or something much more. Sometimes she has disappeared, right from the back seat of the car. If some careless driver fails to pick her up his car may develop, very suddenly, a great deal of trouble. Drivers who have given an old woman a ride from Hilo to Keaau may suddenly see her again the same day in Kona, many miles away from where they last saw her. On an island where most people are known by sight, an old woman may appear whom no one has ever seen before, but who looks vaguely familiar. Sometimes, rarely, Pele shows herself as the strikingly beautiful goddess, only to vanish and leave one wondering if he actually saw her at all.

The stories about her are endless, about her appetites and jealousies, her trysts and her terrors. Modern legends have developed into a folklore of their own: a bottle of gin has largely replaced 'ohelo berries as an offering to the goddess; one does not drive over the Nuuanu Pali after midnight with pork in the car; the small white dog seen high on the slopes of the volcanoes, where no dog could hope to survive for long, is Pele's pet.

An ancient chant could speak wonderingly of Pele . . . .

"O, the passing of that beautiful woman.
Silent are the voices on the plain . . . ."

A temptress then and a temptress now, she is a *presence*, infusing the very matter of that raw, beautiful and windswept world of the volcanoes. And all men who venture into her world pay the price—to be forever haunted by the grandeur, the beauty, and the spirituality of the high volcanoes.

LADY IN RED making offering to
Pele at Aloi Crater. Cinder cone in
background is prehistoric Puu
Huluhulu. At right, the fire goddess
Pele is recreated in a painting by
Howard Hitchcock.

THE PLACE

*"Till when the windy mountain ridge
Buds with the rosy petals of dawn.
Here stand I to wait her relenting."*

"The rich verdant hue of these fairy parks was relieved and varied by the splendid carmine tassels of the ohia tree. Nothing was lacking but the fairies themselves . . . ."

The fascinating landscapes of Hawaii's volcano country inspired Mark Twain, and charmed him. Small wonder. From the distant jumping-off point of Hilo, thirty-one miles away, the volcano country begins to rise to dramatic vistas too vast, too awesome, to see quickly or forget easily. The vegetation at sea level, lush and fecund, gradually gives way to hardier plants and then to fern and ohia trees. Then the trees disappear and the lunar landscape of the high volcanoes captures the imagination with strange colors and shapes. The ground beneath one's feet evolves from the paved streets and ordered gardens upward through wild and lovely forests to end on high and windswept peaks nearly fourteen thousand feet high and often draped in snow.

Five volcanoes—Kohala, Hualalai, Mauna Kea, Mauna Loa and Kilauea—built the island of Hawaii, largest of the hundred-odd islands of the Hawaiian Archipelago. It is the latter three, and Kilauea in particular, that men today call "volcano country," and are drawn to again and again. Around Kilauea, in particular, visitors are attracted to the rich variety of views, from forests to rocky deserts in the space of a few miles' drive.

*STEAM VENTING from floor of
Kilauea caldera.*

11

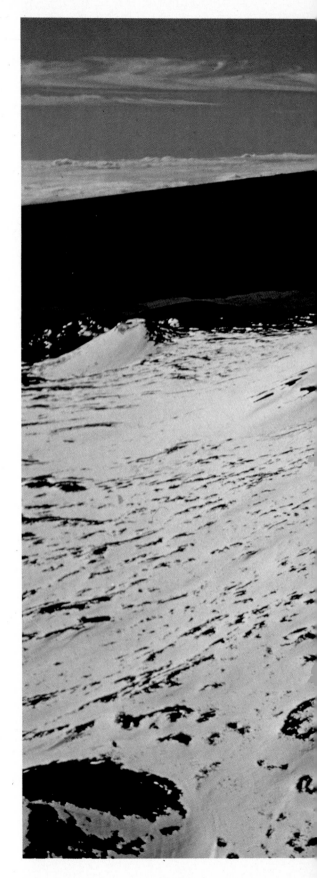

TWO THOUSAND YEARS AGO, 13,967-foot Mauna Kea (in foreground) erupted. Now dormant, Mauna Kea and her sister volcano, Mauna Loa (in background) are snow covered for nearly five months of the year. Mauna Loa is the largest volcano on earth, and has erupted frequently during historic times—its lava covering over half of the Big Island of Hawaii.

The green, growing areas have a universal and timeless appeal; they are the flowering rain forests of startling colors, where strange birds' cries fall musically in a calm morning. Flowers and plants are astonishing in their form and shape. The longer reaches of the countryside contain wild goats and pigs. In the evenings the forests become dark and mysterious and full of mist and fog.

Nearby, the great caldera of Kilauea contains its own special type of beauty. From Volcano House, a charming and historic hotel near the edge of the caldera, visitors look down into this huge hole at the summit of Kilauea. It is nearly four hundred feet deep and encircles twenty-six hundred acres. On the floor of the caldera is a second pit, Halemaumau. A summit eruption of Kilauea generally means an eruption in Halemaumau and this sets in motion a series of other actions: older Hawaiians come with 'ohelo berries and younger ones come with gin and money, all to be cast into the pit in homage to Pele. Volcano House fills rapidly with people, and the telephone rings constantly because people on other Hawaiian islands seek information on the eruption. Airlines schedule special flights from Honolulu, two hundred miles away. Geologists from the Hawaiian Volcanoes Observatory prepare for sleepless nights and round-the-clock sampling of gases and rocks, of picture-taking and recording their observations. National Park Service Rangers prepare for the tides of visitors, setting up barricades in the more dangerous places. Newsmen rush to the scene and devotees of Pele appear in red velvet dresses to stand at the edge of the firepit.

For to see an eruption is to see the genesis of the surrounding land. To see its

aftermath is to marvel at what has been created. The basaltic building blocks of lava are thrown up in strange and marvelous shapes, to be covered eventually by trees and grasses. The high, alp-like meadows, rain-washed and incredibly green, had their origins scores of miles below the surface.

With every outpouring of lava, new hills and ridges, new peaks and valleys are created, sometimes in a more or less predictable line along the general direction of the rift zone, at other times breaking out in a surprising new location. In places eruptions cut the roads leaving hillocks where a highway used to be. Cracks from pre-eruption earthquakes have made a spider's web of the surface. In spite of these natural disruptions, nature has been benevolent in the area and there are small park-like *kipukas*—oases of green in the middle of later lava flows—which are sanctuaries for rare birds and wild pigs.

The pungent, often damp, vivid green forests do not prepare the visitor for the

startling change that is moments away. The fiery belching of Kilauea volcano has devastated some sectors and the landscape undergoes changes that are sometimes stark, sometimes subtle. A bend in the road, a few extra feet of altitude brings the visitor to a bleak and lava-ravaged stretch, an expanse of pumice and ash, a region where dead-white trees stand like pencil-marks, their limbs stripped by falling volcanic ash. Hard by stand the trees which just escaped the fallout, and they are green and blossoming boundaries to the ashen trees.

In the desert areas and in the caldera of Kilauea the wind kicks up small cyclones of lava dust, and carries the steam from heated ground water upward, blowing the white wisps into tatters of mist before they disappear in the cooler air. It is a place of contrasts, of cinder cones domed with yellowish sulphur, and the black paths of old lava flows streaked in dark relief against the lighter hillsides. Roads have been carved through the area but there

MIST SHROUDED RAIN FORESTS are home to stands of Kahili ginger (far left), one of three ornamental species of ginger which has become established in Hawaii Volcanoes National Park. Above, the brilliantly plumed 'Iiwi feeds on insects and a variety of flowers including the endemic lobelias.

are long miles of largely untouched forests and deserts. Sometimes the smell of sulphur comes sharp and strong, but mostly there is a clean smell—and a scent of adventure.

Twain, who could admire the beauty, could be eloquent about the more barren places:

"We came upon a long dreary desert of black, swollen, twisted, corrugated billows of lava—blank and dismal desolation! Stony hillocks heaved up, all seamed with cracked wrinkles and broken open from center to circumference in a dozen places, as if from an explosion beneath. There had been terrible commotion here once, when these dead waves were seething fire; but now all was motionless and silent—it was a petrified sea!

"The narrow spaces between the upheavals were partly filled with volcanic sand, and through it we plodded laboriously. The invincible ohia struggled for a footing even in this desert waste, and achieved it—towering above the billows

here and there, with trunks flattened like spears of grass in the crevices from which they sprang."

And finally: "We came at last to torn and ragged deserts of scorched and blistered lava—to plains and patches of dull gray ashes—to the summit of the mountain, and these tokens warned us that we were nearing the palace of the dread goddess Pele, the crater of Kilauea."

But the lava works mosaics in the rocks, and the streaks of ancient lava flows make vari-colored stripes in the hillsides. There is a charm in the isolation, an appeal in the uncrowded and undulating hills which are bare except for the boulders ejected in old eruptions. What some men shun as a desert, others explore in a quest for peace and beauty; the volcano area is as pure as a mirror, and it reflects the visitor himself.

Towering over Kilauea is Mauna Loa, "the long mountain," a perfect example of the shield volcano. It is probably the most massive mountain in the world. Across a twenty-mile ocean of lava is a companion,

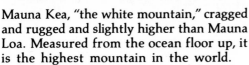
GHOSTLY STAND of dead trees contrasts with the brilliantly flowered Ohia-lehua tree. Above, the Hawaiian short-eared owl, Pueo.

Mauna Kea, "the white mountain," cragged and rugged and slightly higher than Mauna Loa. Measured from the ocean floor up, it is the highest mountain in the world.

These two enormous masses have inspired words and picture images for scores of years. Few words have matched those of the Rev. Titus Coan, describing a trek into this high, crisp area:

"Early the next morning, we pursued our way up the stream, and at noon found ourselves fairly out of the forest with the lofty summit of Mauna Kea rising in hoary grandeur before us. We were now at its base, and in the high, open country occupied by herds of wild cattle. We now bent our course south-south-west, over a beautiful rolling country, sprinkled here and there with clumps of low, spreading trees, which looked like orchards in the distance. Our way was along the upper skirts of the forest, having Mauna Kea with its numerous peaks and lateral craters on our right. At evening we came in full view of Mauna Loa, bearing south by west from us. We pitched our tent under an ancient crater four hundred feet high, now covered with trees and grass.

"Here we had a splendid view of the great terminal crater on the summit of the mountain, about twenty-five miles distant, and also of the vast flood of lava which had flowed down the northern side of the mountain to the plains below, some part of which lay burning at our feet, at the distance of four or five miles. We were now seven or eight thousand feet above the level of the sea; and we could see the dark clouds gather, and the lightning blaze below us, while the deep toned thunder rolled at our feet. At the same time, a storm of hail spread along the shore and fell upon the station at Hilo. This was the first hail seen at our station since our arrival at the Islands.

"At twilight a smart shock of an earthquake, which lasted thirty seconds, added to the sublimity of the scene; while a blazing comet hung over us in the vaulted sky. As darkness gathered around us the lurid fires of the volcano began to glow with fervid heat, and to gleam upon us from the foot of Mauna Kea, over all the plain between the two mountains, and up the side of Mauna Loa to its snow-crowned summit, exhibiting the appearance of vast and innumerable furnaces, burning with intense vehemence, and throwing out a terrible radiance in all directions. During the night we had thunder and lightning; and in the morning both mountains were beautifully mantled in snow."

EARTHQUAKE!

"A boom, as of thunder, from this cliff
A faint distant moaning from that cliff
The island quakes with thy tremor"

To experience an earthquake—particularly one which precedes an eruption—is to know a heart-stopping moment. Professor Charles H. Hitchcock, quoting his friend F.S. Lyman, tells of the one in South Kona, Hawaii, a century ago:

"First the earth swayed to and fro, north and south, then east and west, round and round; then up and down in every imaginable direction for several minutes everything crashing around us; the trees thrashing about as if torn by a rushing mighty wind. It was impossible to stand; we had to sit on the ground, bracing with hands and feet to keep from rolling over." Moments later, the Professor wrote, . . . "there occurred . . . a slide where earth, trees, houses, cattle, horses, goats and men were swallowed up and rocks thrown high into the air."

THUNDEROUS ROCKFALLS and landslides create dust clouds along the cliffs of Puu Kapukapu at the exact moment of a major earthquake. Above, a landslump ruptures Chain of Craters Road in Volcanoes National Park.

ERUPTION!

"The god is at work in the hills
She has fired the plain oven-hot"

"So we climbed for a long time in the plane. At night, to protect your vision the lights are red inside the cockpit. I turned off the blinking lights outside the plane . . . we were flying in a little red cocoon. All of a sudden I noticed the cocoon was extended beyond the airplane . . . and getting pinker. The clouds beyond were getting pinker and I couldn't understand what was happening. Did I have the lights on outside? What's happening? I was on top of pink clouds. Normally at night in the moonlight the clouds are white. In front of me was Mauna Loa . . . the whole mountain was on fire! And wow! It's overpowering . . . here is this mountain on fire . . . we were the only human beings up there, a pri-

vate eruption . . . here's Mauna Loa erupting for the first time in twenty-five years.
Wow! There was a curtain of fire . . . low
fountains, cascading into every major pit
crater. What do you say? What moments
in my life brought me to this?"

Geologist Jack Lockwood, scientist; his
breathtaking view of the Mauna Loa eruption of July 5, 1975, was one shared only
by his companion, Robin Holcomb. It was
the first eruption of the awesome volcano
since 1950, and it was to last for eighteen
hours before subsiding.

After the eruption there was a tremendous flurry of earthquakes, making the
area around the summit very dangerous.
Lockwood flew up again a day or so after
the eruption and found, to his surprise, a
group of people staying at the rest cabin
near the crater of Mokuaweoweo. Two
people were in front of the cabin and he
saw a third person, a woman, on the porch
of the cabin. Knowing the danger they
could face, he radioed the Park Service,
who arranged a helicopter, noting it would
take two lifts to bring out the three people.

Later, Lockwood found the helicopter had made one trip and brought out two persons who said, on questioning, that they were the only ones at the cabin, that there was no woman with them. Still later, Lockwood talked with his companion that day, who confirmed that he had also glimpsed a woman on the porch, a woman wearing a dress, a woman whose face and hair could not be recalled. But definitely a female form on that porch . . . .

Lockwood does not look for human manifestations of Pele. "To me, the manifestation is what's running out of the ground, what's happening." He also contends that someday, cold scientific logic may explain it all away, but for the moment there is an energy force present in the mountain called Mauna Loa that is beyond his capabilities to explain.

Man has marveled at the volcanoes since the beginning of written history. One of the more prolific writers was the Rev. Titus Coan, called the "Bishop of Kilauea" for his intense interest in volcanoes. In a letter written from the port city of Hilo, to the Rev. C.S. Lyman, March 6, 1852, he described an experience:

"At half-past three on the morning of the 17th ult., a small beacon-light was discovered on the summit of Mauna Loa. At first it appeared like a solitary star resting on the apex of the mountain. In a few moments its light increased and shone like a rising moon. Seamen keeping watch on deck in our port exclaimed,'What is that? The moon is rising in the West!'

"In fifteen minutes the problem was solved. A flood of fire burst out of the mountain and soon began to flow in a brilliant current down its northern slope. It was from the same point and it flowed in the same line as the great eruption which I visited in March, 1843.

"In a short time, immense columns of burning lava shot up heavenward to the height of three hundred or four hundred feet, flooding the summit of the mountain with light, and gilding the firmament with its radiance. Streams of light came pouring down the mountain, flashing through our windows, and lighting up our apart-

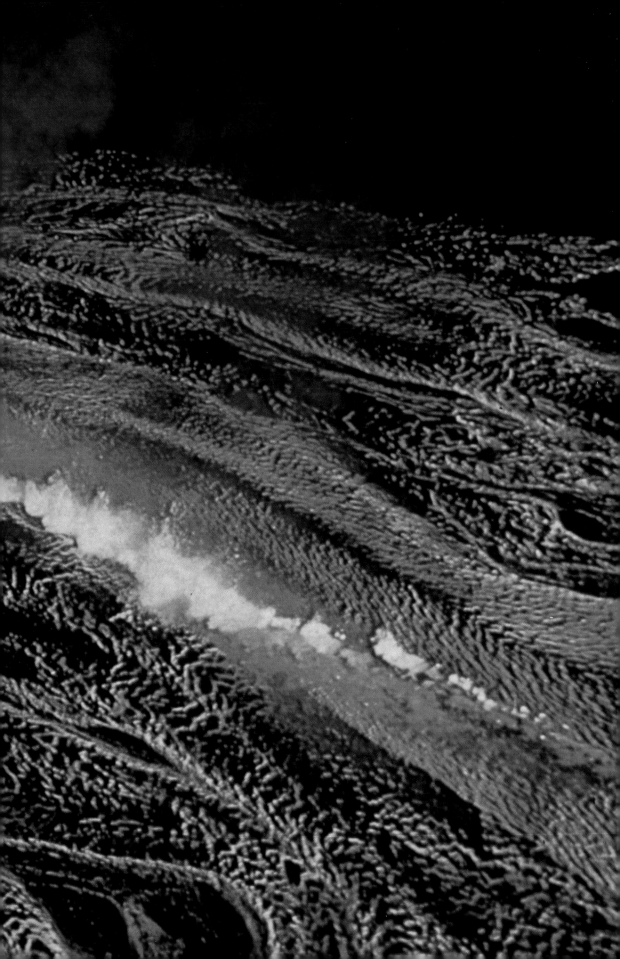

ments so that we could see to read large print. When we first awoke, so dazzling was the glare on our windows, that we supposed some building near us must be on fire; but as the light shone directly upon our couch and into our faces we soon perceived its cause. In two hours the molten stream had rolled about fifteen miles down the side of the mountain."

For the Rev. Coan, for Lockwood, for all the scientists, newsmen, visitors, ancient Hawaiians—the eruptions were and are the most visible movement of lava. The unseen movements take place well underneath the earth's hard surface. For under the calm and green exterior of the island chain, the lava—magma, in its underground form—moves in raw and fluid power. It insinuates itself in fissures in the earth, sometimes creating its own tunnels by the

SUPERHEATED STEAM billows from Mauna Ulu after November 29, 1975, quake. Above, windswept fire fountain and fumes are captured in this time exposure of Kilauea Iki. Wreath of tiny red dots over crater are lights of circling aircraft.

intensity of its size and strength, and under propulsion from sub-surface gases. While the islands sleep, the magma moves upward, coming to a point about two miles below the surface, where it ponds, filling a "reservoir" and roils under pressures until, once again, it begins its upward thrust.

Often, in the early morning darkness the magma reaches the surface and splits the earth. Lava fountains spring up and begin hurling molten rock into the star-bright sky, and the glow is seen for miles.

As the fountains pump the rock skyward, the falling lava cascades back into the fissure and begins to dam up the rent in the earth, forcing the pumping lava into a concentrated fountain of fire which leaps hundreds of feet into the air and fills the countryside with a roar as elemental as that of a jungle beast.

And potentially as dangerous.

Eruptions have killed in the past.

In 1790 the Hawaiian Islands were ruled by regional kings and warfare was a way of life. One of those kings, Kamehameha, was then at war with Keoua, his rival. During their conflict Keoua's army had stopped to rest near Kilauea. As was customary, Keoua paid homage to the fire goddess by making offerings.

In the midst of the rituals, the army was stunned by booming explosions (extremely rare for Kilauea). Some soldiers were killed by the falling rocks. The terrified Keoua tried to make additional offerings to appease the fire goddess, but the explosions became more intense.

It was time for action. To reduce the probability of losing the entire army, Keoua divided the soldiers into three groups. Each group was sent off at intervals along a trail leading to the safety of the Ka'u desert.

The third group impatiently waited their turn and upon command hastened down the trail. To their surprise they soon came upon the second group of soldiers who appeared to be resting near the menacing caldera. But when they walked up to the reclining group, they found to their horror that the soldiers were all dead. There were no burn marks or injuries on the bodies.

Amid the mourning there was curiosity. What had killed their friends? Scientists today believe the army group was caught in a sudden, swift flow of poisonous gases released from the vents with the violent explosions.

Nearly two hundred years later a visitor walks the same trail in comparative safety, knowing the rarity of such explosions at Kilauea.

And yet . . . .

Perhaps the ghosts of the ancient Hawaiians walk as well, for there is a certain unease, a realization that all hell could break loose right underfoot. It is a familiar feeling for volcano-watchers who rush to the eruptions; they hurry to the scene and stay as long as the eruption continues, but they do not kid themselves about the potential for disaster.

Behind the barriers rigged by the Park rangers, visitors are secure, even if the activity were to increase dramatically. For those who must go beyond the barriers—scientists, reporters, rangers, sometimes Red Cross workers and other volunteers—the surging, churning fountain of fire could shift its direction or increase its intensity and put them in immediate and pressing danger. Ash and pumice fall everywhere, wind-driven into eyes and mouths; lava bombs crash chillingly close, and over it all is that terrible noise, like a thousand freight trains, filling the ears and the mind.

When the eruption takes place in a forest

LOW FOUNTAINS PULSE in the circulating lava lake of Halemaumau Fire Pit. The escaping gas fumes (largely sulphur dioxide, water vapor, and carbon dioxide) provided the force that moved the magma to the surface.

on the flanks of Kilauea the hazards increase. Trees burn and fall. Lava flows snake through the forest, threatening to surround the unwary. Sometimes the flows are hard to locate, and there is the knowledge that one could be cut off and not know it, until too late. Off in the brush the sound of moving lava, the crackling of the flaming trees, cause panic which must be fought down. And penetrating the sound and the spectacle is the odor of destruction, the biting, burning smell of sulphur. It clings to the mouth and dries the throat. Days after the eruption has ended the memory of the taste is strong and insistent.

FOUNTAINS OF FIRE

"Puna's dim distant hills are burning—
A glancing of torches—rainbow colors"

On a quiet November night the earth opened up at a site called Kilauea Iki (Little Kilauea) and lava splattered into the pages of history with what would be the highest fire fountain ever recorded—nineteen hundred feet.

It also plunged a young Black Star photographer into an adventure he would never forget.

Naturally the eruption drew an immediate tide of people from Honolulu, and quickly among them were Robert Wenkam and Robert Goodman, both photographers.

Goodman was enchanted with the idea of getting to a real eruption, then fearful he would not get there at all. Even the special flights were full, forcing both to wait anxiously for cancellations.

Finally arriving at Kilauea Iki, they discovered they were the first photographers on the scene, and that five of the initial twelve lava fountains were still in action.

"We were afraid the show might be ending," Goodman said later, "so we decided to go down the switchback trail into the crater a little way.

"Every fifty feet brought us new camera angles. Whenever we stopped to think, we were sure we were too close, but then one of us said 'a little farther.' The thing was irresistible.

"By fits and starts we worked our way

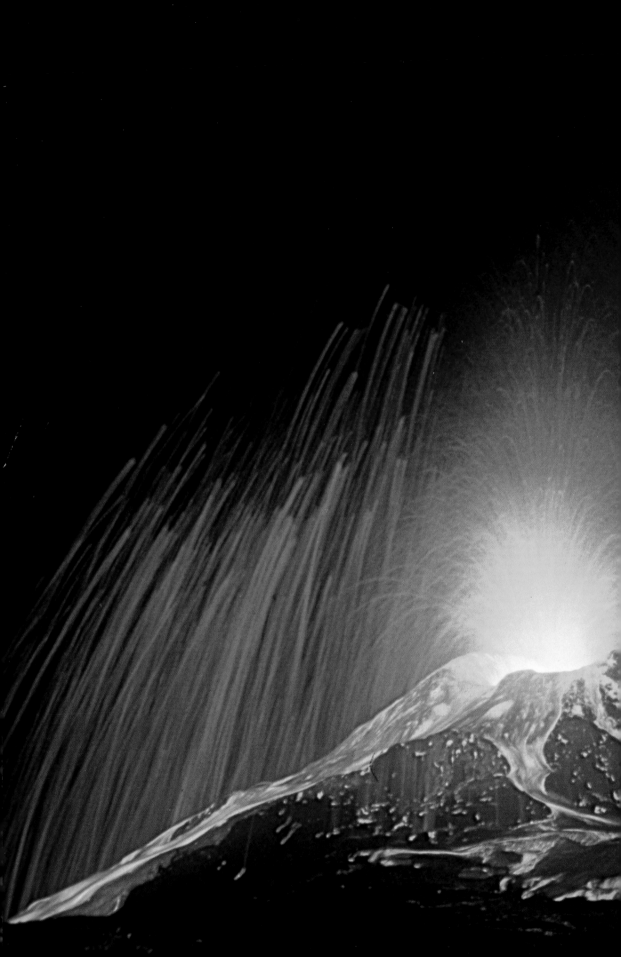

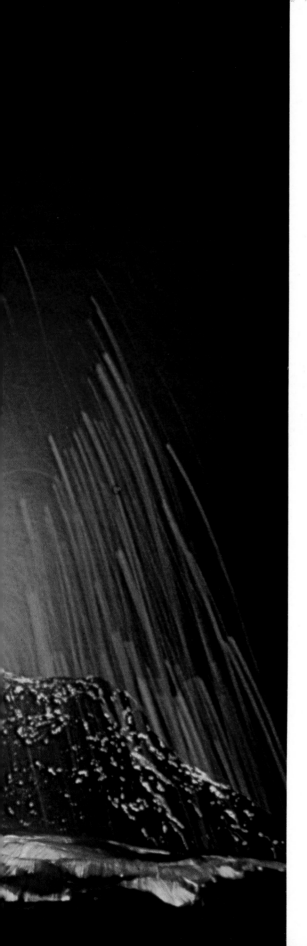

to the rift line. I didn't know we were there until I realized it had suddenly become much hotter. Then I heard a gurgling below me and looked down into the very throat of a fountain that had been playing lava minutes before. I was terrified but I knew this was a chance that might never come again. And dawn was already breaking.

"Soon we were so close to the main fountains that we couldn't stand the heat for more than seconds. We would dart toward it, make exposures, and run back. Each time our clothes seemed to sear us wherever they touched, as if we would burst into flames ourselves.

"The sulphur fumes finally got so bad, our throats felt as if we had swallowed gravel. Daylight had come. We were ex-

BLASTS OF GAS (at left) expel clots of cooling lava from the top of a 60-foot high Mauna Ulu spatter cone. Spatter trajectories are caught here by a time exposure. Above, a Mauna Ulu lava fountain in mid-afternoon.
Underexposure reveals internal structure of fountain. Orange interior is hot, rising column of lava. Darker blobs are cooling lava.

IN THE HALF LIGHT OF DAWN, Kilauea Iki (at right) spews lava to heights of 400 to 600 feet. Above, Puu Kia'i fountains above Kalapana.

hausted. So we worked our way back to the trail, shooting pictures as we retreated. All of a sudden I got scared. For the first time it occurred to me: nobody in the world knows we're down here."

The two photographers packed their equipment and, lungs aching, had scrambled halfway up the trail when they met a team of scientists on the way down. New material. Once more, they started down. Slipping and sliding along a crumbling trail of old lava, they descended to a spot near the floor of the crater where the lava was pooling. The scientists began unpacking their equipment—pyrometers, flasks, long tubes for sampling the lava.

It was a scene out of Dante; all about were burning ohia trees and wet fern plants suddenly crackling and flaming from the intense heat. The roar of the lava fountaining was like a score of locomotives venting their steam in unison, and ash swirled in dark dustdevils, infiltrating eyes and camera lenses. Underfoot was the twisting path and beyond that the old lava flows with their serrated surfaces. Over it all lay the malevolent, blistering heat, an oppressive blanket, worsening as the fire sucked oxygen out of the air.

Then the ordeal was over and they were back on the trail and well above the eruption. Going for provisions—sweaters, hard-hats, water, more film—they raced back to the eruption.

"Once in the crater we felt again the lure to get closer," Goodman said. "But this time the fountain was playing twice as high and seemed twice as hot. We could hear the cinder marching toward us as it splattered on the leaves of the trees. We moved to a ledge overlooking the mouth of the fountain. I felt a shaking and thought it was in my knees. At that point we realized the ledge we were standing on was shaking violently. So we got out of there, and rightly too—that ledge disappeared.

"Finally we ran out of film. Bone-tired and wet by the mist, we found a warm ledge and took a nap, our first in days."

The result of that adventure was a cover story in *National Geographic*, running under a comment that noted, "Never have such elemental displays been better covered by color photography."

*LIKE A GIANT AMOEBA, lava
from Kilauea Iki surges north from vent
area. Lava from this eruption completely
filled the crater area.*

LAVA FALLS

"Here Pele comes from her fortress,
her Mount,
Deserting her resting place, her hearth—
A wild raid down to Malama."

From the Memorial of the Rev. Titus Coan, to Professor J.D. Dana, October 15, 1855:

"In a few days we may be called to announce the painful fact that our beauteous Hilo is no more . . . a flood of burning ruin approaches us. Devouring fires are near us."

Hilo town survived, but what the Rev. Coan described as a "fiery sword" continues to hang over the city. By the simple laws of gravity, all lowland areas are potential targets of a streaming lava flow born in the higher regions. To the people of Pahoa, in the Puna region, the threat is very real. To the people of Kalapana, also in Puna, the lava has been too close for comfort. For the citizens of Kapoho, a small Puna farming community, the threat became a reality, and the village was destroyed by the second phase of the Kilauea Iki eruption of late 1959 and early 1960.

Along with the possibilities of instant disaster, there are moments of great beauty that occur at one time or another during an eruption, or in its aftermath. Such times come in the vari-colored fire fountains themselves, in the olivines (peridot, when polished into gems) that the lava creates, in the wisps of Pele's Hair, in the shapes sculpted by swift lava flowing around and molding against trees.

There is beauty, too, in the lava falls that pour from a high eruption, the long ribbon of red and orange and crusting black matter that flows and drops, inches along and then cascades over a cliff. The author of this book has known many such moments, but one in particular.

It was midnight, and very cold in Kamuela despite a calendar which showed midsummer. Restless, I went into the kitchen and arranged a drink, leaving the lights off. From my window in daylight I could see three volcanoes—Mauna Kea, Mauna Loa and Hualalai, but at night I could only guess at their outlines. Nevertheless, always drawn by them, I stood looking out the window.

I saw a glow begin high in the sky. Blinking at it, I knew in one instant it was not the lights of an airplane, nor a brush fire. In the next instant I felt a thrill jolt through me like electricity. Mauna Loa had not erupted in twenty-five years, but I had seen it begin, and there it was, glowing and getting brighter. It was a summit eruption. My heart leaped.

Soon I was in the Jeep and racing up the Saddle Road toward the eruptions, hastening to beat the barricades I knew would be going up on the lone road between Mauna Kea and Mauna Loa. I turned off on the Mauna Kea road and drove to the nine thousand foot level and parked and sat there all night watching the falls of lava stream down distant Mauna Loa, watched in a bitter wind and under stars that were as brittle as icicles. It was grand. It was beautiful and rare. And if Pele had walked out of the darkness to stand before me I would not have been afraid. Or surprised.

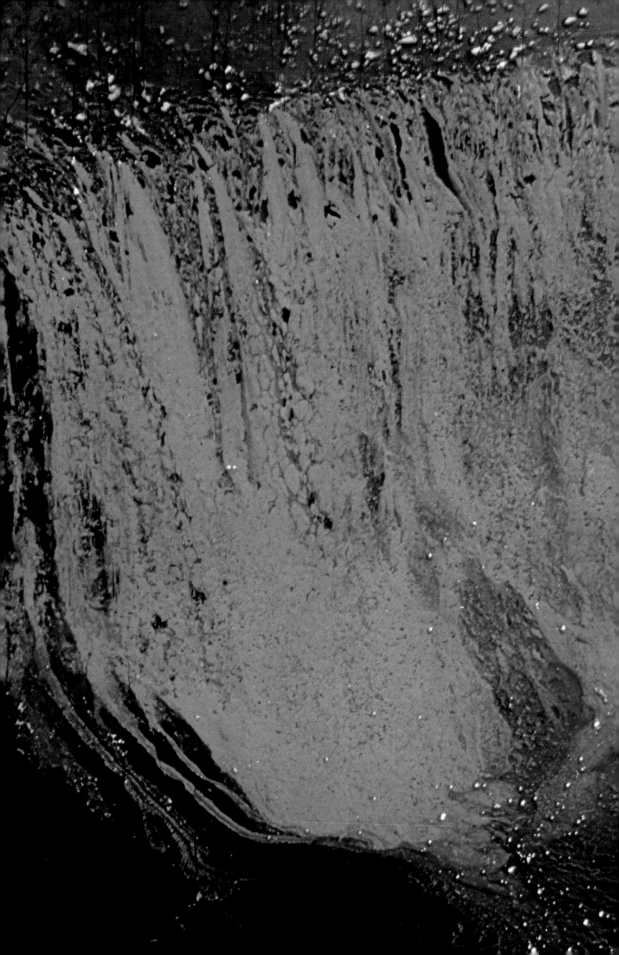

PATTERNS AND COLORS

"And gaunt the forms that jag the sky—
The skeleton woods that loom on high"

The late Reggie Ho of Pahoa, Hawaii, was one of the foremost volcano-watchers of all time. Taciturn and patient, he was known to camp in a place he felt would be the site of an eruption and stay there for days. Sometimes he was wrong, but he was right often enough to do it most of his adult life.

When he was not occupied with other jobs around Pahoa, Reggie was a photographer, and technicians and cameramen around a great deal of the world knew his work. Footage he shot turned up in Hollywood films and on television network news. For when the volcano erupted, Reggie dropped everything else and picked up his battered camera gear and went to the scene. Driven by forces as elemental as the eruption itself, he usually beat everyone else to the fountains of fire. He liked being there first.

He also liked to tell people about the night he outsmarted himself, shaking his head as if to say, "Ah, the uncertainties of life, the foibles of man!"

Reggie was covering a volcanic eruption for a television network. He had shot considerable footage but needed more. With patience and skill he calculated the best time and place to get the magnificent colors and the intricate webs of lava against the Hawaiian countryside. Then, while others

UNUSUAL LAVA POND on west flank of Mauna Ulu shows no gas activity and little surface crust. Bottom, small, highly viscous, arching fountain of lava, early phase Mauna Ulu.

were shooting from ground level, Reggie and a companion worked their way at dusk up a treacherous hillside, with its sliding rocks and uncertain footing, to a small ledge on a hill overlooking the fountain and the flow. It was a dramatic and colorful sight. From the ledge he would be able to photograph the fountain, then follow the flow all the way down to the doomed village and beyond, to the open sea.

Reggie was very pleased with himself. Throughout a good bit of the night he talked about the footage he would get, the exposure, the light. He had decided to wait until the following dawn for the precise moment when the light would be at its most stunning to show the drama and excitement of the scene. That it meant all night on a cold and windswept ledge meant nothing to him. The night was very dark and they could see the flares of trees catching fire, like the striking of a giant match. Underneath the crusting black lava they could see the fluid fire spreading in the rainforest. Just before dawn Reggie got ready to shoot.

He sat the camera in just the right place and made a final check of his gear. Only

then did he discover he was out of film.

His face impassive, he glanced down at the spectacle he might have filmed. Then he began to pick up his gear. He hoisted his tripod and turned to his stunned and speechless companion.

"If you miss it," Reggie said calmly, "you miss it." Then he turned and started making his way down the hill.

Drawn like moths to a gigantic flame, the famous, near-famous and unknowns all have shared moments of awe that were born in the splendor of an eruption. Mark Twain, who found the eruptions irresistible, wrote of the fantastic colors of a lava flow:

"Sometimes (lava) streams twenty or thirty feet wide flowed from the holes to some distance without dividing—and through the opera-glasses we could see that they ran down small, steep hills and were genuine cataracts of fire, white at their source, but soon cooling and turning to the richest red, grained with alternate lines of black and gold.

LIKE LIGHTNING in a summer sky, fire blazes through cracks in the surface of a molten lava lake. Below, a Hornito (after the Spanish word for little oven) belches lava.

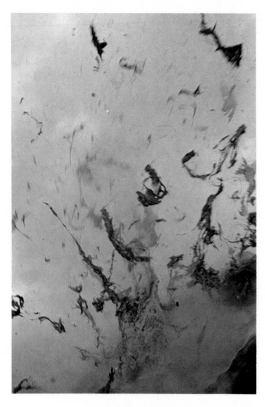

"Every now and then masses of the dark crust broke away and floated slowly down these streams like rafts down a river. Occasionally the molten lava flowing under the superincumbent crust broke through—split by a dazzling streak, from five hundred to a thousand feet long, like a sudden flash of lightning, and then acre after acre of the cold lava parted into fragments, turned up edgewise like cakes of ice when a great river breaks up, plunged downward and were swallowed up in the crimson cauldron.

"Then the wide expanse of the 'thaw' maintained a ruddy glow for a while, but shortly cooled and became black and level again. During a 'thaw' every dismembered cake was marked by a glittering white border which was superbly shaded inwards by aurora borealis rays, which were a flaming yellow where they joined the white border, and from thence toward their points tapered into glowing crimson, then into a rich, pale carmine, and finally into a faint blush that held its own a moment and then dimmed and turned black.

"Some of the streams preferred to mingle together in a tangle of fantastic circles, and then they looked something like the confusion of ropes one sees on a ship's deck when she had just taken in sail and dropped anchor—provided one can imagine those ropes on fire."

More than a century later a geologist would record on tape his own colorful impressions while walking near an eruption: "... there's a lot of blue sulphur dioxide rich gas coming, especially north of the main vent itself ... ejecta coming out right now is not frothy, fumacious spatter that was being shot up a half hour or twenty minutes ago. We're now seeing taffy-like viscous material, and big, elongate blobs, small bathtub size globs that hang together ...."

SHREDDED LAVA is hurled skyward with great force in this shoreline explosion. Seawater entering a new lava tube vaporized instantly. Bottom, cooling pieces of lava emerge from the exploding steam cloud. Right, the surface tension of cooler lava forms a rare dome-shaped fountain at Mauna Ulu.

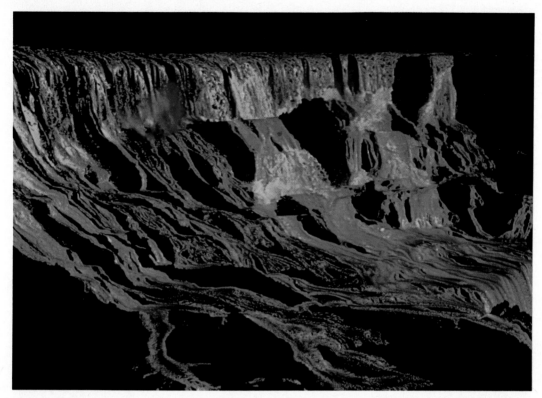

RIVER OF RAPIDLY FLOWING pahoehoe lava streams downward from the throat of Kilauea Iki volcano. Above, lava cascades into Aloi Crater during the twelfth phase of the Mauna Ulu eruption.

RIVERS OF FIRE

"The fire-split rocks bombard the sun
The fires roll on to the Puna sea
The meaning of this wild vision?
The meaning is desolation."

"A large fissure opening through the lower rim of the crater gave vent to the molten flood which constantly poured out of the orifice, and rolled down the mountain in a deep, broad river, at the rate, probably, of ten miles an hour. This fiery stream we could trace all the way down the mountain, until it was hidden from the eye by its windings in the forest—a distance of some thirty miles. The stream shone with great brilliancy in the night, and a long horizontal drapery of light hung over its whole course.

"We gained a little eminence in the woods, from which we could see the lava stream which was now opposite us on our left, distant six miles. This fiery flood was now half way through the forest, and more than three-fourths of the way from the crater to the shore, sweeping all before it."

The Rev. Titus Coan's view of the lava rivers in this instance was that of detached observer. It is a rare view, for the flows are at once ominous and magnificent. They carry both the aura of birth and the scent of destruction.

A LAVA TUBE FORMS when the surface of a flowing channel of lava (at left) cools and crusts over. Lava is molten basaltic rock, and when cooled becomes an excellent thermal insulator. Because of this the lava beneath the crust is able to continue flowing protected by the very lava tube it has created. At bottom right, stalactites in a lava tube ring the edge of a collapse, or "skylight," in the tube roof. Combustion of hot volcanic gases as they encountered atmospheric oxygen may have superheated the rock around the skylight causing it to melt partially and drip to form the stalactites.

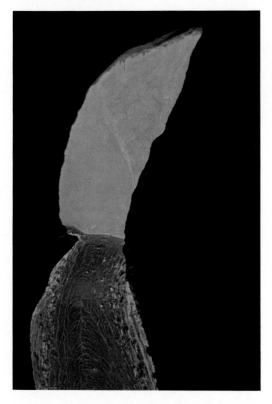

feet, we saw a vast tunnel of subterranean canal, lined with smooth vitrified matter, and forming the channel of a river of fire, which swept down the steep side of the mountain with amazing velocity.

"The sight of this covered aqueduct—or, if I may be allowed to coin a word, pyroduct—filled with mineral fusion, and flowing under our feet at the rate of twenty miles an hour, was truly startling. One glance at the fearful spectacle was worth a journey of a thousand miles.

"We gazed on the scene with a kind of ecstacy, knowing that we had been traveling for hours over this river of fire, and crossing and recrossing it at numerous

Later, and much closer to the event, Coan wrote about the spectacle of a flowing river of lava with power and feeling:

"The lava on which we were treading gave . . . evidence of powerful . . . action below, as it was hot and full of seams, from which smoke and gas were escaping. But we soon came to an opening . . . twenty yards long and ten wide, through which we looked, and at the depth of fifty

points. As we passed up the mountain, we found several similar openings into this canal, through which we cast large stones; these, instead of sinking into the viscid mass, were borne instantly out of our sight upon its burning bosom.

"Mounds, ridges, and cones were also thrown up along the line of the lava stream, from the latter of which, steam, gases, and hot stones were ejected into the air with terrible hissings and belchings."

GLOWING FIERCELY, below, lava flows from vent of 1977 Puu Kia'i (Hill of the Guardian) eruption. At right, engulfing everything in its path, a pahoehoe flow sweeps down the southwest rift of Kilauea.

"With sure and solemn progress the glowing fusion advances through the dark forest and the dense jungle in our rear, cutting down ancient trees of enormous growth and sweeping away all vegetable life. For sixty-five days the great summit furnace on Mauna Loa has been in awful blast. Floods of burning destruction have swept wildly and widely over the top and down the sides of the mountain. The wrathful stream has overcome every obstacle, winding its fiery way from its high source to the bases of the everlasting hills, spreading in a molten sea over the plains, penetrating the ancient forests, driving the bellowing herds, the wild goats and the affrighted birds before its lurid glare, leaving nothing but ebon blackness and smouldering ruin in its track."

MASSIVE AA LAVA FLOW
floods down the north flank of Mauna Loa's upper east rift in July of 1975.

The fascination for flowing lava persists today, but much of the emotion is in the eye of the beholder. The unthreatened observers look and see the beauty and sweep of the incandescent rivers; farmers whose papaya and orchids and coffee are being burned alive view the flows with bitterness.

On January 14, 1960, an eruption broke out on the east rift zone of Kilauea. In time, the voluminous lava formed a fan-shaped flow that covered an area of four square miles. Spellbound observers watched it cover much agricultural land and destroy the village of Kapoho and continue an inexorable movement to the sea. In spite of the destruction, in spite of the dangers, men gathered to watch the eruption until it stopped on February 19th—gazing all the while, as the Rev. Coan phrased it, "with a kind of ecstacy."

A MOVING MOUNTAIN of incandescent lava crunches slowly toward the town of Kalapana. Miraculously, the 40-foot high aa flow stopped just short of the town.

DOOMED VILLAGE of Kapoho with
fountain in background, January,
1960. At left, lava moves down the
Chain of Craters Road above Naulu
picnic grounds.

Y. NAKAMURA STORE

EVACUATE THE TOWN!

"All about is flame
The coco-palms are gone, all gone
Clean down to Ka-poho"

The village of Kapoho lay under the "fiery sword" for years, slumbering in the Puna sunshine, or awash in the bountiful rain. Around it seeds exploded in the fecund earth and great farms of orchid and papaya groves brought prosperity to the villagers.

But only weeks before, the spectacular eruption of Kilauea Iki had stirred the underground magma, and now earthquakes were shattering the peaceful countryside.

53

Worried villagers began to gather their families close, and await developments.

Scientists and newsmen arrived on the scene. The ground was shaking from time to time, but the danger did not appear imminent or extreme. National Guardsmen and trucks were standing by.

As the day wore on it became clear a decision was going to be necessary. It was a consensus decision, finally, and the word was: evacuate the town.

The villagers, knowing it was inevitable, helped load their belongings on the trucks. It was a sad convoy that took them out of the village where most of them had spent their lives.

That night a curtain of fire erupted in a papaya grove, eventually to seal itself off into one fountain which reached fifteen hundred feet. The eruption went on for more than a month. It destroyed more than seventy structures, ground over the orchids and papaya, filled a natural springs resort area with thirty feet of lava, buried a highway and burned a school. It swallowed in fire a group of resort homes along the beach, and at last plunged into the ocean, sending a steam cloud high into the clear Hawaiian skies.

The fiery sword had struck.

MASSIVE LAVA FLOW ignites Kapoho School, while at right, falling volcanic cinder strips leaves of Kapoho papaya tree grove.

SCIENTISTS AT THE EDGE

*"The rocks melt away in thy flame
Fierce rages the Pu'u-lena
Home of the gods, inviolate"*

Outside the aircraft the temperature was below freezing. The altimeter showed fourteen thousand feet, three thousand above the lava flow. It looked hot, felt hot despite the altitude, and the scientist flying the small yellow plane was worried by a strange smell.

"I looked up and saw fluid dripping off the wing," geologist Jack Lockwood recalled later. "No sweat," he said, "there's still water dripping off the wings. It's still wet. And all at once I realized—wow!—that's not water, that's paint. My paint was blistering and melting.

"I looked at the wings closely and saw little black spots beginning to form. Scorch spots! Like a child with a magnifying glass, the lava flow was focusing its heat on our fabric covered wings. We turned as fast as you can safely turn at 14,000 feet and headed for Hilo. When we were within five hundred feet of the runway, the scientist with me calmly asked about the exposed tires on the plane. For all I knew we had a bunch of fried bananas there. I made the best landing I've ever made in my life. There was no problem with the tires, but there were holes in the wings, some charred fabric, blistered wood."

It was the kind of heart-stopping adventure that has become almost commonplace with the scientists and scholars who stand at the edge of creation. From their vantage point of knowledge, they are able to observe the phases of birth, the genesis of the land.

The Hawaiian Volcano Observatory is a part of the volcano investigation program of the U.S. Geological Survey; it attracts some of the best young scientists in the nation to what is a living laboratory for volcano studies.

One of the great rewards they mention again and again is that it is not some arcane science they are studying, nor is it for future generations. The energy forces they relate to in their daily lives are highly visible and often unpredictable. The result of their gas sampling, dome measuring, and other pursuits has been a body of knowledge on which Islanders can draw. The scientists become involved in local communities by lending their expertise and their advice on the search for geothermal power to Civil Defense workers, Red Cross people, newsmen and others involved in the eruption cycle. It is, as one of them put it, a useful kind of science, and applicable now.

You see them at all hours, carrying instruments and devices and pushing off by helicopter or four-wheel-drive vehicles for hidden corners of the volcano country and it becomes clear that while they are adding to the body of knowledge about volcanoes— an admirable enough occupation—they also are having a lot of fun.

Sometimes their scientific approach is challenged, and equalled, by folklore. Several years ago there was a woman in the Puna District, where a lot of eruptions have taken place, who was said to be able to feel earthquakes so mild they were unnoticed by anyone else. She was particularly adept at feeling them while sitting down.

RESEARCHER GARY WEESNER places electrodes into lava to measure its electrical resistance. This knowledge helps in the development of techniques for predicting the movement of magma underground and thus the possible early prediction of eruptive activity.

With a lot of tremors taking place, the woman was driven down to a lonely road in Puna by several scientists and newsmen, who wanted to prove or disprove her abilities. An accommodating, pleasant woman, she agreed to sit in the center of the road and "record" earthquakes.

It was a clear, moonlit night, and the sea was a ribbon of silver five or six miles away. A slight tradewind from the northeast rustled the leaves of trees and whispered through the rows of sugar cane.

The slightly plump, middle-aged housewife sat in the center of the asphalt road, while the scientists produced a chronometer and pencil and paper. Some newsman found a deck of cards in one of the cars and built a little pyramid of them not far from where the lady sat.

After a while she looked up and said, "There's one." A scientist checked the chronometer and wrote the time in his notebook. The cards had not fallen, nor even trembled. No one else in the group felt a thing.

Again she said, "There's one," and the time was logged. In the course of perhaps an hour she noted at least a half-dozen earthquakes which no one else could feel. Finally, when the wind had blown the cards down and the intervals between her "recordings" grew longer, everyone decided to quit for the evening.

Back at the Observatory, the scientists checked the times against the record kept by the seismograph. The times were almost identical; only once did she mention an earthquake that was not recorded on the seismograph—and by that time, no one was betting on the machine.

Sometimes the scientists enjoy a kind of normalcy; sometimes they are stretched thin, as on November 29, 1975. On that day the worst earthquake to strike the Island of Hawaii in more than a century was centered off the southeastern coast, registering 7.2 on the Richter scale. It was quickly followed by an eruption of Kilauea and a tsunami (seismic sea wave). The opportunities for data-gathering kept the scientists on round-the-clock schedules.

The relationship between the questing scientists and the residents in and around the volcano area has been a success story for years. It is as if, in the presence of such prodigious energies as the volcanoes possess, mere man has been able to put aside his pettiness.

LIKE A SILVERY SPACEMAN, Geologist Wayne Ault in heat reflecting suit collects lava sample from floor of Kilauea Iki. Samples are analyzed for their chemical and mineral composition. From this information, scientists can tell the source area of the lava and how long it has been in the magma chamber below Kilauea.

MARINE BIOLOGIST RICHARD GRIGG
studies lava emerging from lava tube on ocean
bottom. Below, a lava stream meets the sea, south
coast, Island of Hawaii.

LAVA IN THE SEA

"The raging waves engulf the steep coast—
The sea Pele turmoiled at Kahiki"

"What Charles Darwin had postulated, we had just been privileged to witness . . ."

That succinct summary came at the end of one of the most thrilling moments that could come to an adventurous oceanographer-diver-photographer-writer, which are a few of the titles which apply to Ricky Grigg.

"It was Charles Darwin," Grigg says, "who first postulated that coral islands were indeed the uppermost remnants of drowned volcanoes; that volcanoes rise up from the sea, form the foundations for coral reefs, and then slowly subside."

Grigg was one of the first two men to see and photograph lava emerging underwater. This is his story:

"One day in the fall of 1973 my telephone rang; it was a colleague, Jim Moore, a marine geologist in Palo Alto. He told me Kilauea was erupting and the volcano was flowing into the ocean and asked if we could meet on the Big Island.

"I agreed. I had been waiting for years to see if I could get underwater photos of lava entering the sea.

"The next morning Jim and I were tossing in a small boat heading for the northeastern tip of the Big Island of Hawaii. Slipping into our wet suits, we clumsily

61

*LAVA POURS INTO THE SEA at
the Kealakomo seashore during the
1973 eruption.*

loaded our cameras into their waterproof housings.

"For two days lava had been coursing down the mountainside. Now, directly in front of us it was flowing into the sea. Clouds of windswept steam and sulphur fumes billowed up from the water's edge, fouling the air. As Pat Tani edged the Boston Whaler closer to the shoreline, Jim and I were both tense. What would it be like, coming face to face with molten lava underwater? No one had ever done that before. And the scene in front of us was not reassuring.

"Jim and I pulled down our masks and, clutching our Nikonos cameras, rolled backwards into the water. It was hot and murky. As we sank I was suddenly aware of a tremendous roaring sound, the sound of lava fracturing as it met the sea. Miraculously, the deeper we went the cooler and clearer the water became. At thirty feet, only fine, sand-like particles of fresh lava clouded our view.

"Edging slowly toward the shoreline we followed the sloping bottom upward. The roar in our ears was now overpowering, and we could feel the vibrations on our chests and in our sinuses.

"Suddenly, directly in front of us and less than ten ten feet away, a lava tube opened up, pouring forth a stream of glowing molten rock. Hissing and vibrating like a living thing, the flow enlarged, forming an elongated pillow-like lava tube right before our eyes. All at once it collapsed, exploding inward on itself with an incredible sound. Jim and I were jolted backwards. For just a moment we hesitated. Then realizing we were unharmed, we moved back to observe.

"For the next thirty minutes we photographed, dodged volcanic debris, and collected samples of the fresh lava. We were so excited we underestimated the danger. In an instant, as if Madame Pele, the volcano goddess, resented our presence, a surge swept Jim against a glowing vent. The lava seared a hole in his wet suit, but miraculously left him unhurt.

"When our film supply was exhausted and our tanks low on air we began the long swim to the boat. Beneath us, in all directions, the once lovely coral reef was now a barren and broken field of lava tubes, fractured boulders, and swirling sand. Fifty years or more would pass before this area again would be fully covered with coral."

DEATH OF A VOLCANO

"The face of the Sun is hot in Puna.
I companioned, it seems, with a god;
I had thought her to be very woman.
Lo and behold, she's a devil!"

In the aftermath of the flames and the rivers of fire, Pele sometimes is judged harshly, but almost never by the residents of the volcano area who live daily with the threat of her fiery touch.

"Pele," they say, "she come, burn my place. Now she go. I build again." Or, "If she comes, she comes. Maybe she take my *place*. I do not think she will hurt *me*."

Caught in the magic and the mystery, they seldom think of building elsewhere. Pele is a part of their lives. So the fires come and the lava destroys, providing the foundation for rebirth, and rebuilding.

The wake of the rushing lava is rich in strange and wonderful things—burning gases and strands of Pele's Hair; gas bubbles in the rocks; tree molds where the

A FRAGILE BUBBLE of basaltic glass formed by volcanic gases.

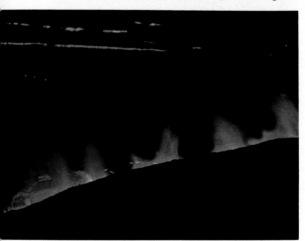

STILL HOT FISSURE is lighted by burning gases in Ka'u Desert.

GLISTENING YELLOW sulphur crystals are deposited by escaping fumes.

rushing lava has flowed around an ohia tree so quickly it burned in place, but left the lava shape; icicles of lava hanging from a rock formed moments ago and still liquid at its core; great "ropes" of fast-moving *pahoehoe* lava; crevices and cracks and jagged ridgelines of *aa*; wild and pleasing lava sculptures, life-size and, in a sense, living.

That beauty has died in the flames is beyond a doubt; that it will live again in the new basaltic earth is richly evident. And despite the cost and the dangers, volcano devotees continue to hope for eruptions, for they are wonderful and awesome, powerful and humbling. They are like something from pre-history, a fiery anachronism in an urban age, reminding us of the raw fury of creation.

ONRUSHING LAVA encased above trees where they stood. At right, burning hydrogen gas lights the tip of a small spatter cone on the Mauna Ulu shield volcano.

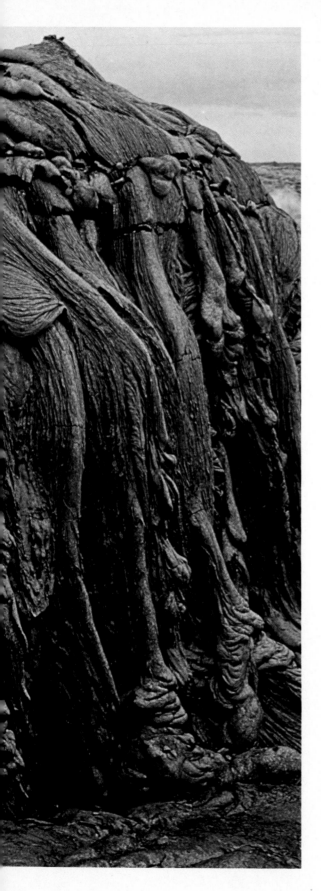

When the flames die and the light disappears, when the smell and the tremors are powerful only in the mind, when the silence suddenly is louder than the noise of a moment ago, then the post-eruption tristesse begins. The air is thick with melancholy. Where there had been life and color and excitement, there is emptiness.

Newsmen mill around as long as possible, then leave one by one. Spectators shuffle off to their cars and are soon gone. Park Rangers remove the traffic signs and barricades and the scientists leave the field to spend the next weeks and months assessing their findings. In a little while the area is quiet.

But deep in the mysterious and unexplored earth, magma is moving. It uncoils like a snake and insinuates itself in small fissures or roars like a mighty wind through the old lava tubes and tunnels. Driven by unimaginable energy, it hammers through the earth and begins to work its way through miles and miles of subterranean passages, climbing toward the surface.

Then, all at once, a glow will be seen for miles and Islanders will look toward Kilauea and Mauna Loa. A curtain of fire will form and with a noise like the birth of thunder a surging fire fountain, wild and lovely, will leap into the night sky.

"The forest-fringe of the pit is aflame
Fire-tongues, fire-globes,
that sway in the wind—
The fierce bitter breath of the Goddess!"

ROPEY PAHOEHOE LAVA drapes
a small cliff-face. At right, an 'ohelo
shrub flourishes in the sheltering arms
of a pahoehoe lava flow. The Hawaiians
considered 'ohelo sacred to Pele, the
volcano goddess.

MAPS

THE HAWAIIAN ISLANDS are the exposed parts of volcanoes that are strung like beads across the ocean floor along a major fracture zone in the earth's crust. The biggest island, Hawaii, consists of five separate volcanoes that have merged to form the highest volcanic pile on earth.

The summit of Mauna Loa towers nearly six miles above its base on the ocean floor.

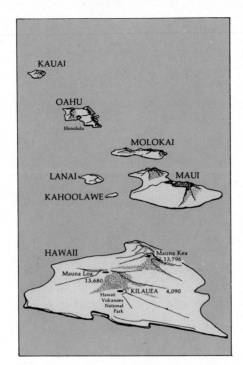

KAUAI

OAHU
Honolulu

MOLOKAI

LANAI
KAHOOLAWE

MAUI

HAWAII
Mauna Kea 13,796
Mauna Loa 13,680
KILAUEA 4,090
Hawaii Volcanoes National Park

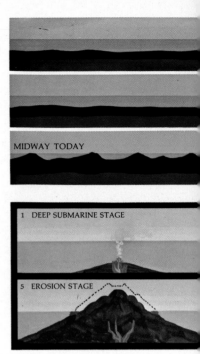

MIDWAY TODAY

1 DEEP SUBMARINE STAGE

5 EROSION STAGE

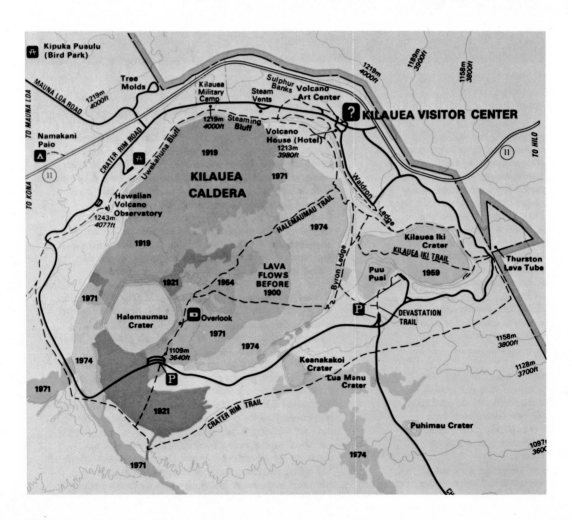

Kipuka Puaulu (Bird Park)

TO MAUNA LOA
MAUNA LOA ROAD

Tree Molds

1219m 4000ft

Namakani Paio

TO KONA
11

CRATER RIM ROAD

Hawaiian Volcano Observatory
1243m 4077ft

Kilauea Military Camp

Steam Vents

Sulphur Banks

Volcano Art Center

1219m 4000ft Steaming Bluff

Uwekahuna Bluff

1919

KILAUEA CALDERA

1971

1919

1921

1954

LAVA FLOWS BEFORE 1900

Overlook

1971

1109m 3640ft

P

1921

1971

CRATER RIM TRAIL

1971

Halemaumau Crater

1974

Volcano House (Hotel)
1213m 3980ft

1188m 3900ft

1158m 3800ft

? KILAUEA VISITOR CENTER

11
TO HILO

HALEMAUMAU TRAIL

Waldron Ledge

Byron Ledge

1974

Puu Puai

1959

Kilauea Iki Crater
KILAUEA IKI TRAIL

Thurston Lava Tube

P

DEVASTATION TRAIL

Keanakakoi Crater

Lua Manu Crater

Puhimau Crater

1974

1158m 3800ft

1128m 3700ft

1097m 3600

70

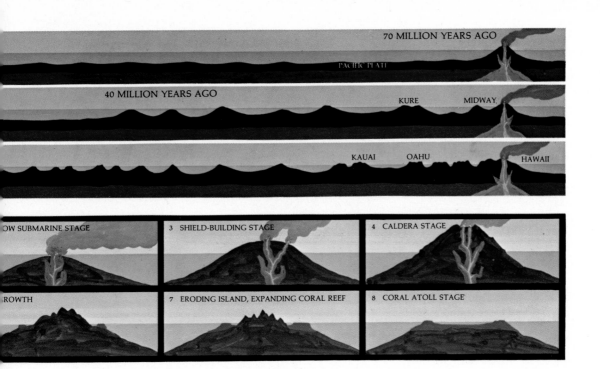

70 MILLION YEARS AGO

PACIFIC PLATE

40 MILLION YEARS AGO

KURE   MIDWAY

KAUAI   OAHU   HAWAII

OW SUBMARINE STAGE

3 SHIELD-BUILDING STAGE

4 CALDERA STAGE

ROWTH

7 ERODING ISLAND, EXPANDING CORAL REEF

8 CORAL ATOLL STAGE

THREE STAGES OF KILAUEA IKI ERUPTION

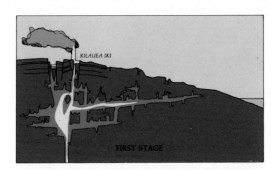

1. Summit eruption at Kilauea Iki in October, 1959 was signaled by earthquakes and swelling of the summit from rising magma. Then, lava fountains erupted along a 1,200 foot rift. The next day, a single strong fountain remained.

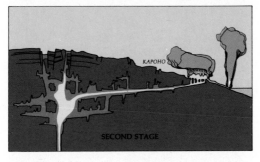

2. Flank eruption at Kapoho followed as seismic tremors continued. Subterranean lava moved out beneath the east rift zone, erupting near Kapoho Village, 28 miles from the summit. Lava from the Kapoho vent covered 4 square miles on its path to the sea.

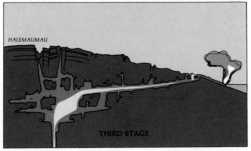

3. Collapse of Halemaumau Crater. Four days after the start of the Kapoho eruption, the summit of Kilauea began to sink as magma drained from beneath the caldera and moved into the east rift zone. Parts of the floor of Halemaumau Crater sank as much as 350 feet.

ACKNOWLEDGEMENTS

This book was produced in association with Hawaiian Airlines, SeaFlite, and Tropical Rent-A-Car Systems, Inc., corporations dedicated to perpetuating the rich cultural heritage of these Hawaiian Islands.

PHOTO CREDITS

pp. 2-3, Robert I. Tilling; p. 4, Jim D. Griggs; p. 5, Robert B. Goodman; pp. 6-7, Don W. Petersen; p. 8, Donald A. Swanson; p. 9, painting by D. Howard Hitchcock; pp. 10-11, Robin T. Holcomb; pp. 12-13, Robert B. Goodman; pp. 14-15, Boone Morrison (2) and Paul Banko; p. 16, Boone Morrison; p. 17, left, Robert B. Goodman, right, Norman Carlson; p. 18, Boone Morrison; p. 19, Donald A. Swanson; pp. 20-21, Don W. Petersen; pp. 22-23, Robin T. Holcomb; p. 24, Boone Morrison; p. 25, Robert B. Goodman; pp. 26-27, Russ Apple; pp. 28-29, Donald A. Swanson; pp. 30-31, Robin T. Holcomb; p. 32, Peter W. Lipman; p. 33, Robert B. Goodman; pp. 34-35, H.V.O.; pp. 36-37, Donald A. Swanson; p. 38, Robert B. Goodman; p. 39, Peter W. Lipman; p. 40, Don W. Petersen; p. 41, Jeffery Judd; p. 44, Robert B. Goodman; p. 45, H.V.O.; pp. 46-47, Don W. Petersen, except upper right p. 47, Glen Kaye; p. 48, Peter W. Lipman; p. 49, Don W. Petersen; p. 50, Don W. Petersen; p. 51, Jim D. Griggs; p. 52, H.V.O.; pp. 53-54, Ray Helbig for Hawaiian Service, Inc.; p. 55, H.V.O.; p. 56, Robin T. Holcomb; p. 57, Larry Katahira, H.A.V.O.; p. 58, Dan Dzurisin; p. 59, Robert B. Goodman; pp. 60-61, Richard Grigg; pp. 62-63, Carl A. Wilmington; pp. 64-65, Robin T. Holcomb; p. 66, from bottom clockwise, Glen Kaye (2), Robert I. Tilling, H.V.O.; p. 67, Robin T. Holcomb; p. 68, Donald A. Swanson; p. 69, Boone Morrison; p. 70, map of islands, Richard P. Wirtz; p. 71, top, Herb Kawainui Kane; bottom, Richard P. Wirtz.